Sidnei Ribeiro Souza

Environmental Management in Metropolitan Regions

Sidnei Ribeiro Souza

Environmental Management in Metropolitan Regions

Analysis of the challenges and legal advances in the Metropolis Statute

ScienciaScripts

Imprint

Any brand names and product names mentioned in this book are subject to trademark, brand or patent protection and are trademarks or registered trademarks of their respective holders. The use of brand names, product names, common names, trade names, product descriptions etc. even without a particular marking in this work is in no way to be construed to mean that such names may be regarded as unrestricted in respect of trademark and brand protection legislation and could thus be used by anyone.

Cover image: www.ingimage.com

This book is a translation from the original published under ISBN 978-620-2-19463-1.

Publisher:
Sciencia Scripts
is a trademark of
Dodo Books Indian Ocean Ltd. and OmniScriptum S.R.L publishing group

120 High Road, East Finchley, London, N2 9ED, United Kingdom
Str. Armeneasca 28/1, office 1, Chisinau MD-2012, Republic of Moldova, Europe
Printed at: see last page
ISBN: 978-620-6-01191-0

I dedicate this work to my family who always supported me.

Acknowledgements

I thank God who through faith sustained me, my supervisor for his commitment and dedication, the course coordinator, the teachers who contributed a lot by sharing their knowledge and friends for their strength.

"Nature is the only book that offers valuable content
in all its leaves." Johann Goethe

SUMMARY

With the approval of Law 13.089 on 12 January 2015, which institutes the Metropolis Statute, the objective of this study lies in the need to understand, comprehend and disseminate it, as well as verify the effectiveness of this instrument in as much as contemporary challenges of environmental issues in metropolitan regions are concerned. To reach this objective, methodological procedures were based on the reading and review of literature correlated to the theme, mainly legal documents, ad hoc consultations with specialists in the field as well as field activities and actions. The results of the work show that the current metropolitan environmental problem reaches complex proportions, where the Metropolis Statute, together with other legal references, makes up important advances, but which still reiterate, in practice, the need for greater interfederative articulation to achieve the precepts of sustainability and prevalence of the common good over the local.

Key-words: Environmental Management, Public Policies, Metropolitan Regions, Environmental Legislation.

SUMMARY

INTRODUCTION

With the consolidation of an increasing number of metropolitan regions in Brazil, a series of problems to the planning and management of the urban and regional environment emerge. This work seeks to analyze the limitations and possibility of Law 13.089 on January 12, 2015, which establishes the Metropolis Statute, for urban development, especially regarding the dimensions of sustainability of cities.

In this sense the work is divided into three parts:

(i)   the first brings a theoretical approach, problematizing environmental degradation in Brazilian metropolises, anthropic effects on natural systems (dimensions and scales of environmental impacts), urban agglomerations, metropolitan regions and the complexity of environmental management;

(ii)   The second refers to the Legal Landmarks that substantiate thinking and acting on the Environment: Forest Code, the 1988 Federal Constitution, the Statute of Cities and the Metropolis Statute.

(iii)   The third and last part properly portrays the parts of the Metropolis Statute, bringing to the discussion its challenges and legal advances, such as the issue of the autonomy of states and municipalities, interfederative governance of natural systems in the metropolis and the potential instruments and limits of urban development by urban environmental planning and management.

Therefore, this work is the result of the concern to understand the environmental legislation along with the growth of cities and metropolises, in addition to the effort to disseminate a legal document that, after several years in Congress and the National Senate, today can provide important public policies for urban sustainability.

In the methodological scope, the work was carried out through a bibliographical review of literature correlated to the theme Estatuto da Metrópole and subjects pertinent to the Metropolitan Region of São Paulo, analysis of legal documents Forestry Code, Cities Statute, Metropolis Statute, Federal Constitution of 1988 among others, visit and consultation of ad hoc specialists and registration through photography of some artificial boundaries between municipalities of the metropolitan region of São Paulo and superposition of data for comparative analysis.

It is expected that this work contributes to highlight the importance of new thoughts and legal alternatives of action on the urban environmental dynamics. It is up to the environmental manager not only the appropriation of these laws, but the analytical and procedural incorporation of its effects on society. In the limit, therefore, a fundamental tool for environmental management of large urban centres, revealing their contradictions, but mainly their potential for development and quality of life.

CHAPTER 1

ENVIRONMENTAL DEGRADATION IN BRAZILIAN METROPOLISES

One of the major problems arising from current urbanization derives from the ever-increasing conurbation of human agglomerations, which is based on the tension generated by the maintenance and political and administrative autonomy of the territories of each federative entity. In other words, with broken territorial limits, whose borders overlap and complex centralities are produced on a metropolitan scale, public policies and legislation itself become difficult to execute and to comply with respectively.

Environmental problems, therefore, move to a regional dimension, and no longer local. As Carvalho (1997, p. 33) emphasizes, "cities reflect, express and reproduce the contradictions of the societies that build them". In the limit, this also leads to reflection on the global dimensions of environmental problems, such as the greenhouse effect, or the contamination of large hydrographic basins. In fact, a certain level of integration becomes increasingly necessary among public policies. However, legal bottlenecks hinder the full implementation of joint actions between the different national entities.

1.1 Anthropic Effects on Natural Systems in the Metropolis

Anthropic effects are the results of human actions on the environment. Planting a tree or cutting it down are anthropic effects, however, the difference will be in what will be the result and the impact that each action will result in the environment. Unfortunately, the anthropic effects with the greatest proportions on our planet are the negative ones. Deforestation, erosion, pollution, burning in preservation areas, expansion of cattle breeding and trafficking in wild animals are some examples of these effects that are increasingly worrying a part of the population that still worries about the preservation of the environment.

In relation to environmental issues, here thought in relation to the anthropic processes that produce them, one can cite numerous phenomena and cases in which these have a direct relationship with the form of agglomeration. In this way, many approaches to socio-environmental phenomena have weighed against metropolization, as this process represents the *anti-nature,* the artificial, the end of the essential contact between the being and the natural entities.

In accordance with CONAMA RESOLUTION N°001, of 23 January 1986, environmental impact is considered to be any alteration in the physical, chemical and biological properties of the environment generated by any form of matter or energy resulting from a human activity, being exercised directly or indirectly, affect:

I - the health, safety and well-being of the population;

II - social and economic activities;

III - the biota;

IV - the aesthetic and sanitary conditions of the environment;

V   - the quality of environmental resources.

Among the most common anthropic effects we have as some examples:

Deforestation: occurs mainly for the expansion of territorial areas for livestock, urbanization and timber extraction. Its effects are the most diverse, with a great environmental impact, since trees are the main means of oxygen "depollution", besides forming the natural habitat of thousands of wild animals. Among the main impacts we can mention: erosion, desertification, reduced soil protection, interference in fauna and flora, contributes to the depletion of water and other natural resources, etc.

Burning: one of the fastest means of clearing land, generally used for the same purposes as deforestation. The impacts are far-reaching, endangering all the local flora and devastating all the fauna of the environment, and the smoke from the act spreads for kilometres, polluting other cities and causing damage to the ozone layer.

Pollution of the soil and rivers/seas: resulting from the negligence of human beings in the disposal of domestic or industrial waste, where the waste is discarded in inappropriate places, such as vacant lots and rivers, which often occurs due to lack of planning when initiating a This act, results in a large amount of impacts. The pollution of rivers and seas causes the death and contamination of fish and other aquatic organisms, proliferation of diseases, floods and even interruption of fishing activities. Soil pollution results in desertification of the site, impoverishment of the soil in nutrients and loss of biodiversity.

Illegal trade in animals and plants: this illegal market also causes damage to the environment, such as extinguishing species, threatening ecological functions that were performed by the species in question, transmission of diseases and stresses in animals that are chased by traffickers, which may even result in the death of the animal.

Brazil is one of the countries with the highest rate of animal trafficking. It represents between 15% and 20% of the world total, which can be estimated in more than 12 million animals taken from their habitat for smuggling. In addition to trafficking to buy these animals for home use, there is the trade in feathers, organs, leather and other parts of the animals.

Introduction of invasive species: often occurs from trafficking in these, can cause impoverishment of biota and natural cycles and threaten sites through predation, competition or alteration of the natural habitat.

Hunting and predatory fishing: Hunting is the pursuit of an animal, usually with the intention of killing it. They generally use ambushes, pursuit at speed and/or group work, always aiming, among the possible targets, at the most fragile, such as old, sick or newborn animals.

Predatory fishing is that which takes more from the environment than it can replenish, reducing the population of fish and even of plants in the ecosystem. Predatory fishing has disastrous consequences and may limit the fisheries productivity, whether from a biological or economic point of view. They use artifacts such as bombs, which are considered of great destructive value, affecting flora and fauna, use of poisons, fishing with fine mesh nets and fishing outside the periods allowed by law.

This action results in the long-term extinction of species, a decrease in the population of organisms present there, which can result in a deregulation of natural cycles where the species was a participant, changes in the ecosystem.

Construction of Dams: we can use as an example the construction of the Serro Azul dam, where its construction implies cutting down 230 hectares of Atlantic Forest.

Urban Expansion: it may be considered one of the worst anthropic effects caused by human beings. This expansion, when it occurs in an irregular way, can lead to serious environmental impacts. Construction of roads, houses and industries, for example, must follow the laws and be supervised by really responsible bodies. When there is a regularity and a correct supervision most of the impacts can be avoided or at least minimized.

This type of anthropic effect can damage the environment in several aspects. It leads to a decrease in local biodiversity, extinction of species, both flora and fauna, soil, water and air pollution, animals being run over, impediment to ecological exchange, reduction of vegetation cover, etc.

1.2 Sustainable Forms of Metropolis

Studies produced in recent years have put to the test the prerogative that the metropolis as a spatial organisation would be negative for relations between society and nature.

In urban and regional planning, authors such as Jabareen (2006), Jenks and Jones (2010) highlight the role that compact and integrated cities have in the new sustainable concepts of urban forms: density, diversity, mixed land use, compactness, transportation, solar radiation incidence and ecological design. As evidenced in the comparative study of the matrices and presented in Table 01, which follows detailed on other recent urban models.

Matrix for analysing sustainable urban forms

Design Concepts (Criteria)	Neotraditional Developmenl	Compact City	Uiban Con lain ment	Eco-City
Density	1. low 2. moderate	1. low 2. moderate	1. low 2. moderate	1. low 2. moderate
	8. High	8. Higti	3. High	8. High
Diversity	1. low 2. moderate	1. low 2. moderate	1. low 2. moderate	1. low 2. moderate
	3. Higli	3. Hiçíi	3. High	3. High
Mixed land use	1. low 2. moderate	1. low 2. moderate	1. low 2. moderate	1. low 2. moderate
	8. High	3. High	3. High	3. High
Compactness	1. low 2. moderate	1. low 2. moderate	1. low 2. moderate	1. low 2. moderate
	3. High	8. Hiçíi	3. High	3. High
Sustainable	1. low 2. moderate	1. low 2. moderate	1. low 2. moderate	1. low 2. moderate

tiansportation	3. High	3. **Hiçli**	3. High	3. **High**
Passive solar design	**1. low** 2. moderate	1. low 2. moderate	1. low 2. moderate	**1. low** 2. moderate
	8. High	3. High	3. High	**3. High**
Greening-	**1. low** 2. **moderate**	1. low 2. moderate	1. low 2. moderate	1. low 2. moderate
Ecological design	3. High	3. High	3. High	**3. High**
Total score	15 points	17 points	12 points	16 points

Note: Scores of the urbon forms are highlighted in bold.

Table 01: Analytical matrix of sustainable urban forms, elaborated by Jabareen (2006).

Sustainable forms associated to compact cities stand out with 17 points and the "Eco-Cities" with 16 points with regard to levels more adequate to socio-environmental situations. This means that, despite the complexity of the new metropolitan territories, the criticism of the population concentration and, in a way, of the environmental factors, may be questionable. Due to the compactness, the buildings become functional due to the proximity and concentration of housing, commerce and services. Environmentally, when there is sufficiently effective planning and management, it favours the reduction of expenses and waste, either by reducing the use of transport, or by reducing urban sprawl, in addition to the concentration of individuals and activities allows more meetings and social cohesion.

In another sense, more than the form of organization, or even the population growth itself, the environmental problems generated in urban spaces, or by large cities, seem to proceed to gaps in planning and management of these territories. In spite of this, the environmental dimension has always been a kind of retardatory approach, or even a synonym of delay or negativity of the modern movement and of the development that was sought. Even in the face of the weakening of this rationality, it is common for inspection agencies of municipal, state and federal governments to be held responsible for the delay in works. This is due to problems with environmental permits, or even disasters sometimes resulting from excessive pressure for the rapid release of large mining, energy, housing and transport projects

1.3 Urban Agglomerations, Metropolitan Regions and the Complexity of Environmental Management

According to Grostein (2001), Brazilian urbanisation resulted in 12 metropolitan regions and 37 non-metropolitan urban agglomerations, which concentrate 47% of the country's population. In 2010, the IBGE indicated a total urban population of 84.3%. According to the cited author, "33.6% of the Brazilian population (52.7 million inhabitants) reside in the 12 metropolitan areas, in extensive conglomerates involving 200 municipalities" (GROSTEIN, 2001). This information is important because it shows that today most of the population resides in cities, which are becoming more and more degraded and incapable of subsidising the quality of life of their inhabitants.

These metropolitan complexes include municipalities with

complementary functions, independent management and unequal financial capacity. These characteristics make it difficult and condition the meeting of social demands and urban infrastructure that, in most cases, arise from the functional relationship between municipalities and depend on solutions that go beyond their political and administrative boundaries, being equated on a regional scale. Metropolitan regions, by contingency or by the nature of the relations established between the municipalities that make them up, would depend on integrated urban development policies and articulated actions, which would be proper of a shared management. The historical absence of such procedures has aggravated the inadequate use and occupation of the soil, with a strong environmental impact. In the 1980s, the peripheries of the nine metropolitan regions grew by 3.1%, while the municipalities at the centre grew by 1.4%. This growth occurred despite the fact that the metropolisation process had slowed down, with the population growth rate dropping from 3.8% in the 1970s to 2%. Even so, the metropolises absorbed 30% of the country's demographic growth in the 1980s, receiving 8.3 million new residents (GROSTEIN, 2001).

A large part of these residents living in favelas, with a 118% growth in the favela population, especially in the regions of Belém, Recife, Curitiba and São Paulo, "in the municipality of São Paulo, 19.80% of the population lives in favelas, on the banks of streams, steep slopes, margins of avenues and under viaducts". (GROSTEIN, 2001). This process results in a great contrast between different metropolitan areas and municipalities, where its unsustainability stands out, as well as the low quality of life due to peripheralisation and environmental degradation.

Attention should be paid to this process, since the sustainability of the urban/metropolitan agglomerate, in its physical-urban component, is related to the following variables: the way of occupying the territory; the availability of inputs for its operation (water availability); the discharge of waste (destination and treatment of sewage and rubbish); the degree of mobility of the population in the urban space (quality of public mass transport); the supply and meeting the needs of the population for housing, social facilities and services; and the quality of public spaces. Thus, the policies that sustain the parcelling, use and occupation of the soil and the urbanistic practices that make these actions feasible play an effective role in the goal of leading the cities along the path of sustained development. [The evolution of this process resulted in the worsening of predatory environmental practices, generating soil erosion, flooding, landslides, deforestation and pollution of water supplies and air, affecting the urban set and especially the areas occupied by low-income population, with significant losses and diseconomies for the proper functioning of the metropolitan set. The strength of the peripheral urbanisation pattern evidenced the negligence of the State, in its different instances, with the construction of cities and the formulation of an urban development policy; illegality as a structural factor in the dynamics of urban expansion of Brazilian metropolises; the precarious urban lot, the house in the slum and the rent of a room in tenements as the predominant alternatives to solve the housing problem of the poor in the metropolis; the absence of a metropolitan housing policy; the insufficient public production of social housing in face of the demand; and the absolute disregard of the society and of the public power with the resulting socio-environmental problems. (GROSTEIN, 2001).

Besides the metropolitan regions, another form of organization for planning and management refers to the urban agglomerations. These, according to Matos (2000), are conceptually new in Brazil, and also refer to population nuclei in the city, that is, even in areas legally defined as rural.

> By extension, it can be assumed that the urban agglomerations, when greatly expanded and exceeding "certain limits and sizes", would conform another territorial unit, the urban agglomeration. This, in turn, is closely associated with the term metropolis, which, in its modern urban sense (disseminated within urban planning as a field of knowledge), refers to the existence of a relatively large urban area comprising more than one municipality, the "metropolitan region". This spatial category presupposes the existence of a main city which organises, economically and functionally, nearby peripheral localities. As a consequence, a dense urban network should emerge where industrial, commercial and service activities are installed, concentrating capital, labor force and political power (MATOS, 2000).

For not having a precise limit of delimitation the IBGE ends up using the terminology in situations in levels below the metropolitan, where a process of expansion of the urban mesh and conurbation of the cities begins. These characteristics are thus framed when the metropolitan region is not established by the State, becoming easy to criticize, since some agglomerations have more articulation, population and other characteristics than some Brazilian metropolitan regions.

One of the dimensions of these problems, arising from the increasing conurbation of human agglomerations, lies in the tension generated by the maintenance and political and administrative autonomy of the territories of each federative entity. In other words, with broken territorial limits, whose borders overlap and complex centralities are produced on a metropolitan scale, public policies and legislation itself become difficult to execute and to enforce respectively.

1.3 Metropolitan Governance and Environmental Management

Environmental problems move to a regional dimension, and no longer local. As Carvalho (1997, p. 33) highlights, "cities reflect, express and reproduce the contradictions of the societies that build them". At the limit, this also leads to reflection on the global dimensions of environmental problems, such as the greenhouse effect, or the contamination of water resources, as well as human degradation in spaces unfit for the development of life in society (Figures 01 and 02). In fact, a certain level of integration becomes increasingly necessary among public policies. However, legal bottlenecks hinder the full implementation of joint actions between the different national entities.

Favelisation in Jardim Itaberaba (western zone) of the city of São Paulo

Figure 1: Irregular occupation in Itaberaba, human-environmental degradation.
Photo: Sidnei Ribeiro, 2017.
Bairro Chácara Maria Trindade in the city of São Paulo

Figure 2: Irregular occupation in risk areas in the Chácara Maria Trindade neighbourhood in the city of São Paulo, on the border of the cities of Santana de Parnaíba and Cajamar.
Photo: Sidnei Ribeiro, 2017.

In the case of Brazil, the metropolises are mostly concentrated on the eastern edge of the territory, set practically less than 100 kilometres from the Atlantic coast. So too, the demographic density that in cities of the RMSP reaches 13 to 15 thousand inhabitants per square kilometre (Diadema and Carapicuíba, respectively), when analysed throughout Brazil, this same indicator does not exceed 25 inhabitants per square kilometre. This reveals the wide agglomerative difference of the national population, as well as the difficulty of managing environmental issues among so many diversified realities (GUZMÁN et. al. 2006).

According to data from the Observatory of the Metropolises (2015), Brazil today has 70 Metropolitan Regions, 3 Integrated Development Regions and 4 Urban Agglomerations.

Number of MRs and year of implementation

1C RM	IT	REGIONS METROPOLITAN	CREATION	N* RM	UF	REGIONS METROPOLITAN	CREATION
1	AL	RM ÀEjeâte	30.11.2009	36	PE	RM Recife	08.06.1973
2	AL	RM of Zona da Mata	15.12.2011	37	PR	RM Apucaiana	12.01.2015
3	AL	RM of Caetss	26.07.2012	38	PR	RM Campo Mourân	12.01.2015
4	AL	RM of Palmeira dos Indios	05.01.2012	39	PR	RM Cascavel	32.01.2015
5	AL	RM of Médio Sertão	08.08.2013	40	PR	RM Curitiba	08.06.1973
6	AL	RM of Sertão	26.07.2012	41	PR	RM of Umuarama	22.08.2012
7	AL	RM of the Paraíba Valley	15.12.2011	42	PR	RM Londrina	17.06.1998
8	AL	RM Maceió	19.11.1998	43	PR	RM Marinsá.	17.07.1998
9	AM	RM Manaus	30.05.2007	44	PR	RM Toledo	12.01.2015
10	AP	MR Macapá	26.02.2003	45	RJ	RM Rio de lameiro	01.07.1974
11	BA	RM Feira de Santana	06.07.2011	46	RN	RM Natal	16.01.1997
12	BA	RM Salvador	08.06.1973	47	RR	RM Centra]	21.12.2007
13	EC	RM Cariri	26.06.2009	48	RR	RM of the Capital	21.12.2007
14	EC	RM Fortaleza	08.06. 1973	49	RR	Southern RM of the State	21.12.2007
15	ES	RM Grande Vitória	18.01.2005	50	RS	RM of Serra Gaucha	30.08.2013
16	GO	RM Goiânia	30.12.1999	51	RS	RM Porto Alegre	08.06.1973
17	MA	RM Greater San Lnis	25.05.2015	52	SC	RM Carboniferous	09.01.2002
18	MA	RM Southeast Maranhão	17.] 1.2005	53	SC	RM Chapeed	20.12.2010
19	MG	RM Belo Horizonte	08.06.1973	54	SC	RM of Alto Vale do Itajai	20.12.2010
20	MG	RM Ajo Valley	04.01.2012	55	SC	RM of Contestado	24.05.2012
21	MT	RM Cuiabá River Ditch	27.05.2009	56	SC	RM Far West	24.05.2012
22	PA	RM Belém	08.06.1973	57	SC	RM FLorianopolis	09.09.2014
23	PA	RM Sanrarém	17.01.2012	58	SC	RM Foz do Rio Itajai	09.01.2002
24	PB	RM Campina Grande	11.12.2009	59	SC	RM Lages	26.01.2010
25	PB	RM of Ararana	21.01.2013	60	SC	North-Northeast RM Catarinense	06.01.1998
26	PB	RM of Barra de Santa Rosa	13.07.2012	61	SC	RM Shark	09.01.2002
27	PB	RM of ÇaiazeuriE	08.06.2012	62	SC	RM Vale do Itajai	06.01.1998
28	PB	RM of Hope	08.06.2012	63	IF	RM Aracaju	29.12.1995
29	PB	RM of Itabaiana	21.012013	64	SP	RMBaÊrada Samtiata	30.07.1996
30	PB	RM de Sousa	21.01.2013	65	SP	RM Campinas	19 06.2000
31	PB	RM of the V. of I'. íamanzuape	21.01.2013	66	SP	RM of Sorocaba	20.06.2014
32	PB	RM Pianeó Valley	06.07.2012	67	SP	V. do Paraíba and Northern Coast MRV	09.01.2012
33	PB	RM Guarabira	12.07.2011	68	SP	RM Sâo Paulo	08.06.1973
34	PB	RM Joâo Pessoa	30.12.2003	69	TO	RM Gunipi	05.04.2014
35	PB	RM Patos	27.12.2011	70	TO	RM Palmas.	08.01.2014

Crp-riTr^itr. Jc number of F.agiòsa Fbnqnliianjz in Brazil. We highlight the high number of mdiopoliianaz regions created in the states of PR (3), AL (S), PB (9) and SC (13\

Source: IBGE. 2015 (Ccmpcsiçào óeEIvís. REDEs e Aglomerações Urbanas). Orgnniraçân do autor. 2016.

Table 02: Metropolitan Regions by 2015, extracted from Oliveira (2016).

As Levy (2015) explains, the geographical boundaries between metropolitan municipalities are artificial borders, where most cities formally and informally send their sewage, waste, air pollution and traffic (Figure 03). This implies in a problematic that in practice is joint, however, in theory still continues to be divided by these political-administrative limits.

Artificial border between the municipalities of Barueri (left) and Jandira (right)

Figure 03: Artificial border between the municipalities of Barueri (left) and Jandira (right).
Photo: Sidnei Ribeiro, 2016.

Being a metropolis in itself does not make it unsustainable. In fact, it is the conditions of this process that will imply difficulties or solutions. Clean energy choices, sewage treatment, reduction of pollutant gas emissions and, above all, integrated actions between the different metropolitan cities together may in the near future improve urban environmental quality and conditions.

CHAPTER 2

LEGAL INSTRUMENTS FOR ENVIRONMENTAL MANAGEMENT IN METROPOLITAN REGIONS

Even with the legal advances, regarding environmental management in Brazil, metropolitan governance is still permeated by a series of disputes and development difficulties, created in diverse contexts and scales, from agglomerations with little over thirty thousand inhabitants, up to twenty million citizens. In the legal support, many devices can be listed and briefly summarized to substantiate a more grounded and qualified action of the public power on environmental impacts in cities and metropolises.

2.1 Urban Development and Environment in the 1988 Federal Constitution and the regulations of Law 10.257/2001

CF-88 advances by bringing a perspective of urban development in line with the social function of urban land and the concern for the well-being of its inhabitants, as advocated in Articles 182 and 183,

Article 182. The objective of the urban development policy, implemented by the municipal public power in accordance with general guidelines established by law, is to order the full development of the social functions of the city and guarantee the   well-being of its inhabitants   .

§ The master plan, approved by the Municipal Council and mandatory for towns with more than twenty thousand inhabitants, is the basic instrument of urban development and expansion policy.

§ 2 - Urban property fulfils its social function when it meets the fundamental requirements for the ordering of the city as expressed in the master plan.

§ 3. Expropriation of urban real estate shall be carried out with prior and fair cash compensation.

§ The municipal public power is entitled, by means of a specific law for an area included in the master plan, to require, under the terms of federal law, the owner of unbuilt, underutilised or unused urban land to promote its appropriate use, under successive penalties of

I - compulsory parcelling or building;

II - tax on urban land and property progressive over time;

III - expropriation with payment through public debt bonds of an issue previously approved by the Federal Senate, with a redemption term of up to ten years, in annual, equal and successive installments, ensuring the actual value of the compensation and legal interest.

Art. 183. Whoever possesses an urban area of up to two hundred and fifty square metres as their own for five years, uninterruptedly and unopposed, using it for their dwelling or that of their family, will acquire dominion over it, provided that they do not own another urban or rural property.

§ 1° - The domain title and concession of use shall be conferred to the man or to the woman, or to both, irrespective of marital status.

These premises allow one to think of environmental quality as one of the elementary factors of urban development, however the effective regulation of the articles, including the guarantee of the right to sustainable cities will only be materialized thirteen years later, with Law 10.257/2001, known as the Statute of the Cities. This law provides for the full implementation of articles 182 and 183 of CF-88, establishing the rules of public order and social interest which regulate the use of urban property for the collective good, safety and well-being of citizens, as well as the **environmental balance**.

Still in the Statute of the Cities, in the guidelines, already in the second paragraph, the need for "planning the development of the cities, the spatial distribution of the population and the economic activities of the Municipality and the territory under its influence area, in order to **avoid and correct the distortions of the urban growth and its negative effects on the environment"** (LAW 10.257/2001, emphasis added). In the same article, still as a guideline, the law provides two more important sections which support the concern with sustainability in the city, starting with the following: "**protection, preservation and recovery of the natural and built environment, of** the cultural, historical, artistic, landscape and archaeological heritage; [...] **hearing of the municipal Public Power and the population interested in** the processes of implementation of undertakings or **activities with potentially negative effects on the natural or built environment**, the comfort or safety of the population". (LAW 10.257/2001, emphasis added). In addition to the centrality of the environmental issue in urban-environmental legislation, in the latter, the question of democratic management is now more effectively posed. This is an important advance, since the dimensions of nature present in the metropolis may, from then on, be required in a regulatory and normative manner, making it become a collective entity and a public decision on its destinies.

In addition to the environmental defense effort instituted by the Constitution, its regulation through the Statute of the Cities allows specific actions, such as filing injunctive actions against individuals and groups that attempt against the environmental balance of the cities, taking there a broader dimension than purely the elements of natural order, but also the transformed environment. In such a way, seeking to guarantee the "**right to sustainable cities**, understood as the right to urban land, housing, environmental sanitation, urban infrastructure, transportation and public services, work and leisure, for the present and future generations (LAW 257/2001, emphasis added).

The revolutionary character of Law 10.257/2001 should be highlighted, as it advances CF-88 and provides the country with modern political instruments for urban development today. It, however, runs into bottlenecks that permeate a broader dimension of planning and management of cities, especially when the municipalities are increasingly connected in the same urban spot, passively or actively sharing the positives and negative effects of the metropolization process. Even to the point of interfederative disputes arise over public services, with protracted judicial proceedings for decisions that are not always consensual on the fate of environmental management. One of the most important cases of these tensions between environmental governance of metropolitan regions can be indicated with the Direct Action of Unconstitutionality (ADI) n° 1.842-RJ, which was filed in the Federal Supreme Court (STF) with the request that articles 1 to 11 of Complementary Law n° 87, of 16 December 19971, articles 8 to 21 of Law n° 2.869 of 18 December 1997, as well as Decree 24.631 of 1998, all normative acts issued by the State of Rio de Janeiro and which legally sustain the institution of the Metropolitan Region of Rio de Janeiro and the micro-region of the Lakes.

This is a case of disputes over sanitation services by federal entities in the metropolitan area, which, among other questions, referred to the contravention of Article 30 of the Federal Constitution of 1988, whose hypothesis resided in the State's intervention over the municipalities' competence to "legislate on local interests", as well as to "organise and provide, directly or under a concession or permission regime, public services of local interest, including collective transport, which is of an essential nature". Among the arguments raised by the plaintiff is the idea that the rules for the creation of the Metropolitan Region by the State would take away competent functions from the municipalities, putting at risk the Brazilian federative pact.

This case, in addition to others, such as ADI 2077-BA, with disputes over public services in metropolitan regions, led to the recent approval of law 13.089/2015, known as the Metropolis Statute. For years in Congress, the law seeks "general guidelines for the planning, management and execution of public functions of common interest in metropolitan regions and urban agglomerations established by the States, general rules on the integrated urban development plan and other instruments of interfederative governance, and criteria for the Union's support for actions involving interfederative governance in the field of urban development, based on items XX of art. 21, IX of art. 23 and I of art. 24, in § 3 of art. 25 and in art. 182 of the Federal Constitution" (BRASIL, 2015).

In addition to the provision of services, many conflicts permeate the border areas of metropolitan municipalities, as well as urban agglomerations, or even Integrated Development Regions, which in the latter case encompass interstate agglomerations in addition to municipal ones. In São Paulo, despite advances in recent decades, these limits are gaps or areas that are more difficult to manage (Map 01), either due to distant access, or due to the environmental characteristics of the presence of closed forests, or steep and irregular terrain, as well as domination of the territory by criminal factions.

SÃO CAETANO DO SUL
SANTO ANDRE
SÃO PAULO
DIADEMA
SÃO BERNARDO - DO CAMPO
Região Metropolitana de São Paulo
Área Urbana
Canal de Drenagem
Projeção Geográfica SIRGAS-2000
1:75.000
0 0,75 1,5 3 4,5 6 Km
MAPA 01: MANCHA URBANA E MEIO AMBIENTE - RMSP, 2017
INSTITUTO FEDERAL DE
EDUCAÇÃO, CIÊNCIA E TECNOLOGIA
SÃO PAULO
Câmpus São Roque

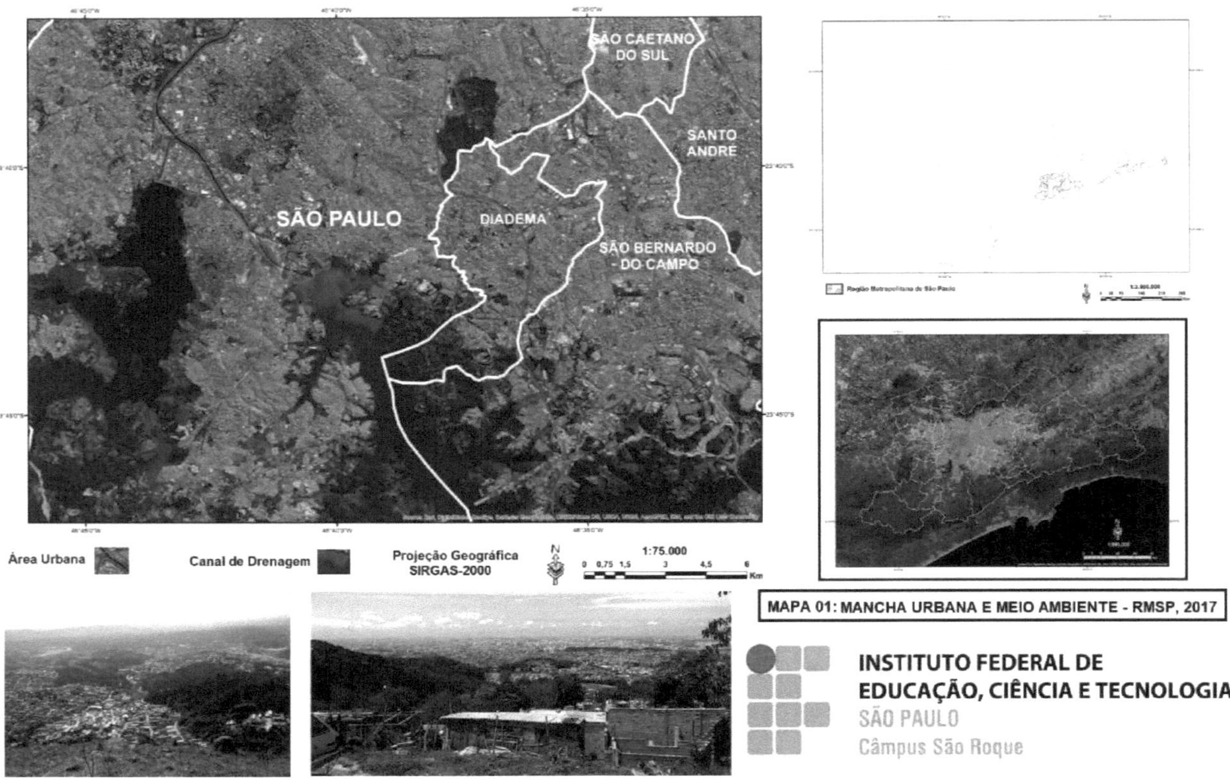

18

As for the problematic of integration of services and management actions among municipalities, some possibilities are regulated with the Statute of Cities, such as Article Art. 34-A, which establishes that "in metropolitan regions or urban agglomerations instituted by state complementary law, interfederative consortium urban operations may be carried out, approved by specific state laws. (Included by Law No. 13.089, 2015). "(LAW 10.257/2001). And which came to be regulated with the Metropolis Statute, which will be further explored in the last chapter.

However, as denounced by Maricato (2002), the arsenal of modern laws, even if coherent, ends up dissipating in an unbalanced practice, without educational actions or an inspection capable of putting these formal and normative mechanisms into practice. In an interview conducted in 2016 with the Secretary of Urban Planning of Osasco, Simone Beralda, the manager explains important advances noted in environmental actions in the scope of the metropolis through EMPLASA,

> With the integrated development plan that is being led by EMPLASA, so to speak, and we believe that it will have a new dimension to work in a cooperative way, because there are many studies made by EMPLASA from the regional point of view. I believe it will give a beautiful leverage in this issue of Inter federative cooperation. (Simone Beralda, Director of Urban Planning of Osasco - RMSP, interview held in the first half of 2016).

In the interview, the director also highlights the difficulties in integrating environmental policies with other municipalities in Greater São Paulo, however highlighting the role of strategic planning and urban environmental management preponderant in the local scale. According to her, there are many environmental problems throughout the RMSP, especially with the APPs and the non-compliance with the relevant legislation, especially due to a culture that devalues the public in reason of the private:

> The most serious problems we have are the disposal of debris on the banks of streams mainly on municipal boundaries which in turn are areas of APP, in public areas and that often is a cultural issue of many years despite the law of Use and Occupation of Soil prohibiting. In the first decades of the city's emancipation, it was very common to receive APP areas in lots while the law prohibited receiving APP areas with slopes greater than 30%.then the municipality would only receive the donation of the APP area with a slope greater than 30%, i.e. areas that could not be destined for the installation of public facilities and could not be occupied because it was an APP. (Simone Beralda, Director of Urban Planning of Osasco - RMSP, interview conducted in the first half of 2016).

The manager also highlights the advances verified in the last years, when a greater effort was made to expand the green areas and to activate the instruments of the City Statute, in order to better plan and manage the urban-metropolitan environment in Osasco. He also criticizes the constant legal changes and the urban dynamics that hinder municipal action, such as the growth of irregular land subdivisions, overlapping municipal and state entities, among other problems.

> As time went by we implanted various parks and planted trees in various areas. But you see what happens when SEMA receives an undertaking that has an environmental impact, it makes the same charge as CETESB. CETESB, when there is an intervention in an APP, the interested party has to ask for the pertinent licenses to act in a certain area, so it is asked to plant trees six times bigger than the area the interested party intends to occupy, But what usually happens is that we donate seedlings of native vegetation but we cannot plant them because public areas have not been reserved for planting. There was a lot of irregular occupation that ended up reducing even more the 0,69 m/2 being further away from the 12m/2 because the amount of green area required was much higher than what the Federal Law 6766 demanded which was 30% and today is 50% of institutional green area and there even so how do you get to such a low volume, It is even nice to do a survey that tells us that today considering all the urban operations that we have plus the implementation of parks that had an increase of 73% of green areas, we know that even so it is still far below the recommended and today to reverse this situation we would have to make expropriations because everything is occupied, is dense and the demand for resources is a very high investment and the regularization of areas
>
> (Simone Beralda, Director of Urban Planning of Osasco - RMSP, interview conducted in the first half of 2016).

Osasco is a municipality located in the West Zone of RMSP, with over 1.5 million inhabitants and whose urban reality is inserted in the limits with other extremely densely populated areas, where the level of conurbation makes the political-administrative division only theoretical (Map 02). In the municipality there is ample favelisation, above all in the limits with Carapicuíba and north of the Tietê River, degraded areas, coexisting with low-income populations, polluting industries and middle- and upper-class condominiums.

Map 02: Sidnei Ribeiro and Rafael Oliveira Organisation, 2017.

Through the approval of Law 13.089/2015, establishing the Metropolis Statute, many of the problems and difficulties of integrated planning and management are legally grounded. However, many practical and even legal obstacles still remain for the sustainable environmental management of the metropolis, as shall be observed in the following chapter, where this legal regulation shall be detailed.

CHAPTER 3

CHALLENGES AND LEGAL ADVANCES IN ESTATUTO DA METRÓPOLE

Law 13.089/2015, called Metropolis Statute, stems from a long public discussion held in recent years by the new demands of the urban-metropolitan phenomenon. It is, very simply, about meeting the precepts of urban development and sustainable cities, required by CF-88 and partially regulated by Law 10.257/2001, being, therefore, one of the paths for practical performance of the environmental manager in today's world. What, however, appears as an alternative, a solution, in practice still faces many challenges, making it justifiable to analyze each point of the Law, its potential and its barriers to the current reality.

3.1 Provisions of the Metropolis Statute (LAW No. 13.089/2015)

According to the Metropolis Statute, interfederative governance follows the following principles; common interest among municipal entities, sharing responsibilities, democratic management of public resources, aiming at **sustainable development in order to minimize environmental impacts regarding** planning, execution, allocation of resources, accountability, shared actions of public functions with the participation of civil society. Therefore, this law

> "[...] establishes general guidelines for the planning, management and execution of public functions of common interest in metropolitan regions and urban agglomerations established by the States, general rules on the integrated urban development plan and other instruments of interfederative governance, and criteria for the Union's support to actions involving interfederative governance in the field of urban development, based on items XX of art. 21, IX of art. 23 and I of art. 24, in § 3 of art. 25 and in art. 182 of the Federal Constitution. (BRASIL, 2015).

The Metropolis Statute also includes the Micro-regions established by the States.

The Law does not solve the problem of absence of parameters for the creation of MRs, AUs and MRs, since each State is responsible for this, but it solves some of their difficulties regarding integrated public services and other points directly associated with environmental issues (sanitation, waste disposal, energy, among others). About this difficulty of delimitation between the different units, Matos (2000) explains the following:

> By extension, it can be assumed that urban agglomerations expanding greatly and exceeding "certain limits and sizes", would conform another territorial unit, the urban agglomeration. This, in turn, is closely associated with the term metropolis, which, in its modern urban sense (disseminated within urban planning as a field of knowledge), refers to the existence of a relatively large urban area comprising more than one municipality, the "metropolitan region". This spatial category presupposes the existence of a main city which organises, economically and functionally, nearby peripheral localities. As a result a dense urban network should emerge where industrial, commercial and service activities are installed, concentrating capital, labour force and political power. The issue of whether space should exceed a certain dimension, reaching another

scope, that of the region, is a controversial point, as there is no consensus
on what the size (minimum or maximum) of this space should be, which
makes it relatively imprecise. To partially circumvent such difficulties the
IBGE has been adopting the concept of "urban agglomerations", which
although similar to that of metropolitan region, serves to designate other
urban spaces, located at sub metropolitan level, which congregate more
than one city, notably cities that would start to experience the process of
conurbation. (MATOS, 2000).

Even not resolving these delimitations (although in the guidelines Law establishes some fundamental characteristics[2]) and according to Levy (2015), with the new law (Metropolis Statute), "the management of common interest of the participating municipalities will be prioritized in the integrated development planning, consortiated urban operations, division of competences. Where it will influence issues such as waste management, sanitation and mobility as these issues isolated from the cities being discussed jointly through interfederative governance." (LEVY, 2015).

In the second paragraph of the legal text, it still stands out, in addition to a directive presentation already exposed, the attempt to define the integrated urban development plan, in addition to metropolitan region and metropolis, highlighting that the criteria for the delimitation of the region of influence of a regional capital, should be considered the goods and services provided by the city to the region, covering industrial products, education, health, banking services, commerce, jobs and other pertinent items, and will be made available by the IBGE in the world computer network. Even so, without necessarily considering qualitative criteria, or compared with broader dimensions of the national territory or with other political-administrative regions of the country. Here, from the outset, a contradiction is evident: that of the attempt to bring classification criteria, but the generality and absence of the multiscale approach still makes the State omnipotent in this delimitation and not the municipalities, or the Union itself.

3.2 The Autonomy of States and Municipalities Faced with the Metropolis Statute

In the scope of resolving impasses between different interests of municipalities in a given metropolitan region, one of the main objectives of Law 13089/2015 lies precisely in the establishment of parameters to solve these conflicts. For, according to Oliveira (2015),

In the planning sphere, the consequences of this process are reflected in
the difficulties created in determining important points for action,
especially when the role and function of each entity is unclear. In a more
practical way, the emergence of conflicts is revealed in the dispute over
housing, public transport, basic sanitation and public security services,
among others. It is no longer possible to precisely define the limits of the
entity's action, and what has always been tacitly its responsibility is now
being called into question and interests are arising on an increasingly
wide scale. Such focus is justified by the criticality of human relations in
the city in relation to nature, especially when it is observed that, even in
the face of technical and scientific advances, as well as the dissemination
of information and the implicit possibilities of cooperation, **less than half
the world's population is served by sewage system, or yet, that the
supply of treated water is not a universalized component in an
expressive global economy.** (OLIVEIRA, 2015, our emphasis)

More specifically in relation to municipal autonomy, established by CF- 88, Oliveira (2015) explains that this is one of the most important points of the text, as

> It implies a broader questioning of the functions of each territorial unit in the country, especially with regard to the sanitation sector, an area in which investment costs are high, making it a natural monopoly activity that requires long-lasting contracts, but which do not always converge with the speed that urban transformations require. Even more when this sector lacks major interventions and expansion of services, since not even half of the Brazilian population is served by sewage network, or still, the growing and constant number of floods could be reduced by simple strategies of drainage control. (OLIVEIRA, 2015).

It seems that even centrally, the Metropolis Statute continues in chapter II, to leave the State with the function of establishing the metropolitan regions and acting centrally on the purposes of common good outlined in the legislation: "Art. 3 **The States, through complementary law, may establish metropolitan regions and urban agglomerations,** constituted by grouping bordering Municipalities, to integrate the organization, planning and execution of public functions of common interest". (BRAZIL, 2015).

In addition, "State and Municipalities included in metropolitan regions or in urban agglomerations formalized and delimited in the form of the caput of this article must promote interfederative governance, without prejudice to other determinations of this Law. This seems flawed, since there is no clarity about these losses, since the councils representing the municipalities are still incipient and the weight of each management unit is not structured. This makes the State, or even the consolidation of the regional unit, to favour unilateral interests in this process. Continuing, there are still cases similar to the RIDES, in which the Statute puts it as follows: "Art. 4º The institution of metropolitan region or urban agglomeration involving municipalities belonging to more than one State will be formalized through the approval of complementary laws by the legislative assemblies of each of the States involved." (BRAZIL, 2015).

Finally, the sole paragraph states that "the approval of the complementary laws foreseen in the **caput of** this article by all the States involved, the metropolitan region or urban agglomeration will be valid only for the

Municipalities of the States that have already approved the respective law". (BRAZIL, 2015). To this end, respecting this process, state laws should define:

> I - the municipalities making up the urban territorial unit;
> II - the functional fields or public functions of common interest that justify the institution of the urban territorial unit;
> III - the conformation of the structure of interfederative governance, including the administrative organization and the integrated system of resource allocation and accountability; and

The text advances in the sense of placing States and Municipalities at the forefront of regional, metropolitan or urban agglomerate regulation, however, it is reinforced, that the absence of parameters for the institution of a council, a collegiate and management system that harmonises the different interests. Difficulty already existing and which seems to remain in the face of the above.

3.3 Interfederative Governance in the Metropolis by Law 13.089/2015

The interfederative governance of the environment is based on important principles for sustainability and should be listed for knowledge and application. These are related to all articles and points, but it is worth highlighting some that are believed to be more central to environmental promotion in the short and medium term:

- **Observance of regional and local peculiarities:** because natural aspects differ within an established political-administrative unit. Attention should be paid to this fact, considering the need to preserve local biodiversities.

- **Democratic management of the city:** everyone should participate in the planning and management of cities, considering social diversity and the role that natural or unnatural assets have for each group or segment of society.

- **Effectiveness in the use of public resources: to** enhance non-economic and productive uses in a rational and balanced way with social demands and the regeneration capacity of the natural system.

- **Pursuit of sustainable development:** thus respecting the economic, social and environmental precepts when making decisions on the expansion and development of cities.

It is in chapter 4 of the Statute that there is an attempt of the general guidelines to seek integrated governance aiming at the principles listed and analysed above. Thus, considering the following guidelines:

I - implementation of a permanent and shared planning and decision-making process for urban development and sectorial policies related to public functions of common interest;
II - establishing shared means of administrative organisation of public functions of common interest;
III - establishment of an integrated system of resource allocation and accountability;
IV - shared execution of public functions of common interest, through cost sharing previously agreed upon within the interfederative governance structure;
V - participation of civil society representatives in the planning and decision-making processes, in monitoring the provision of services and in carrying out works related to public functions of common interest;
VI - Making multi-year plans, budget guidelines laws and annual budgets of the entities involved in interfederative governance compatible;
VII - compensation for environmental or other services provided by the Municipality to the urban territorial unit, in accordance with the law and agreements signed within the framework of the interfederative governance structure.
Sole Paragraph. In applying the guidelines established in this article, the specificities of the municipalities integrating the urban territorial unit as to population, income, territory and environmental characteristics shall be considered. (BRASIL, 2015 - our emphasis).

And it is Article 8, Chapter III, which establishes the basic governance structure, placing the representation of the executive power, a collegiate instance, public consultative organisation and technical support capacity and, finally, an integrated system of integrated allocation of resources and accountability. This last element, a loophole for structuring an urban development fund, considering that the Metropolis Statute was approved with vetoes in this article referring to the establishment of a joint fund.

Complementing the broad range of legal instruments for the promotion of urban development in Law 10.257/2001, the Metropolis Statute, in Chapter IV, Article 9, specifies other fundamental elements of the interfederative governance plan:

I - integrated urban development plan;
II - interfederative sectoral plans;
III - public funds;
IV - consortiated inter-federative urban operations;
V - zones for the shared application of urbanistic instruments provided for in Law no.° _10.257, of 10 July 2001;
VI - public consortia, in compliance with Law No.° 11.107, of April 6, 2005;
VII - cooperation agreements;
VIII - management contracts;
IX - compensation for environmental services or other services provided by the Municipality to the urban territorial unit, according to item VII of the **head of** Article 7° of this Law;
X - interfederative public-private partnerships (BRASIL, 2015)

Whereas these assumptions should compose an integrated urban development plan, renewed every 10 years, covering urban and territorial areas, of which should be established, at least:

I - the guidelines for public functions of common interest, including strategic projects and priority actions for investment;
II - the macro-zoning of the urban territorial unit;
III - the guidelines as to the articulation of the Municipalities in the parcelling, use and occupation of the urban soil;
IV - guidelines on the intersectoral articulation of public policies affecting the urban territorial unit;
V - the delimitation of areas with urbanization restrictions aimed at protecting the environmental or cultural heritage, as well as areas subject to special control due to the risk of natural disasters, if any; and
VI - the system for monitoring and controlling its provisions. (BRAZIL, 2015).

In the case of the RMSP, a large part of these activities are already developed through EMPLASA, a public company, responsible for ensuring the management and planning of metropolitan regions in São Paulo and which, with the publication of the Statute, now has greater legal arsenal for its actions. With the publication of the Statute, it now has greater legal arsenal for its actions.

Finally, in chapter V, the role of the Union emerges, which will be to support the initiatives of the other entities, considering other laws and legal assumptions. This is relevant considering its role of acting in poorer areas, or those where there is greater technical and organizational difficulty in establishing development plans. Likewise, from a restrictive posture of deliberate creation, without any criteria, of metropolitan regions, the latter, very precariously established in Article 15, where the Union will make understanding as AU in the cases where the agglomeration does not fit in the initial basic precepts of MR. Finally, the support for integrated development by the Union should be based on the following terms:

Art. 13 - In its actions included in the national urban development policy, the Union shall support the initiatives of the States and Municipalities aimed at interfederative governance, with due regard for the guidelines and objectives of the multi-year plan, the goals and priorities established by the budget guidelines laws and the limits of available resources provided by the annual budget laws.
Art. 14 For Union support to interfederative governance in metropolitan regions or urban agglomerations, it shall be required that the urban territorial unit has full management, under the terms of subitem III of the **head of** Art. 2° of this Law.
§ 1° In addition to the provisions set forth in the **head of** this article, Union support for interfederative governance in metropolitan regions requires compliance with item VII of the **head of** Article 2° of this Law.
§ 2° The Union may support the preparation and review of the integrated urban development plan pursuant to Articles 10 to 12 of this Law.
§ 3° Additional requirements for Union support to interfederative governance will be established by regulation, as well as for the micro-regions and cities referred to in § 1° of art. 1° of this Law, and for public consortia formed to operate in public functions of common interest in the field of urban development.

Art. 15 - A metropolitan region established by supplementary state law that does not comply with the provisions of item VII in the **caput of Art.** 2° of this Law shall be classified as an urban agglomeration for the purposes of public policies under the responsibility of the Federal Government, regardless of whether or not the actions in this regard involve the transfer of financial resources.
Art. 16 The Union shall maintain actions aimed at the integration between twin cities located on the border strip with other countries, in relation to urban mobility, as provided for in Law no.° _12.587, of 3 January 2012, and other public policies related to urban development.

3.4   Pondering the Challenges and Advances of Law 13.089/2015 in Urban Environmental Management

In addition to regulating the possibilities of interfederative governance of large Brazilian urban nuclei, the Metropolis Statute establishes several guidelines and orientations for Integrated Urban Development, touching on important points for sustainable development. Some of these, in addition to those already mentioned, are inserted in the creation of an integrated multivariate information system, thus establishing a technical and functional support bank for development actions, as specified in the final provisions of the legal text:

Art. 20 - The enforcement of the provisions of this Law shall be coordinated by the public entities that make up the National Urban Development System - SNDU, ensuring the participation of civil society.
§ 1° The SNDU shall include a subsystem of metropolitan planning and information, coordinated by the Union and with the participation of state and municipal governments, in the form of regulation.
§ 2° The metropolitan planning and information subsystem shall gather statistical, cartographic, environmental, geological and other data relevant for planning, management and execution of public functions of common interest in metropolitan regions and urban agglomerations.
§ 3 The information referred to in § 2 of this article should preferably be georeferenced.

In force since January 12, 2015, the Metropolis Statute was approved with important vetoes, such as the creation of a fund and without necessarily establishing practical and important criteria for the creation of metropolitan regions.

Thus, it also runs the risk of being one more modern and important legal document that is little applied. However, it establishes the fundamental bases for urban planning and management, and this includes the environmental dimension in its entirety, from social aspects to natural systems. It is thus a breakthrough in the search for sustainable development in order to minimize environmental impacts in terms of planning, execution, allocation of resources, accountability, shared actions of public functions with civil society participation in planning and decision-making in the implementation of municipal budget laws.

Especially in those where conurbation has increased over the decades and problems have arisen, among them those of an environmental nature, such as the generation and disposal of waste, increased flow of motor vehicles aggravating traffic congestion and causing increased air pollution, suppression of vegetation, pollution and increased contamination of water bodies, use and occupation of the soil without proper urban planning, because these problems have dragged on for decades bringing disruption affecting the quality of life of the populations of these regions.

CHAPTER 4

CONCLUDING REMARKS

We intend through the use of different research methodologies, such as bibliographic reviews pertinent to the theme, analysis of legal documents, superposition of statistical data, qualified semi-structured interviews, to verify in a succinct and concise way the contributions of the aforementioned law, analysing its guidelines that may guide the management and public policies through interfederative governance, aiming at the common good of the federative entities, especially the municipal ones.

The metropolis besides having its socio-economic importance can be sustainable, some authors emphasize the compact cities as sustainable because the concentration of people centralized near their jobs and the diversity of trade optimize and minimize the displacement reducing transport costs and consequently the reduction of emissions of pollutants from vehicles whether from public transport or private and consequent centralization of public services reducing costs saving the public purse.

After researching and briefly analyzing the problems arising from the process of urban densification in metropolitan regions, especially in the Metropolitan Region of São Paulo, the sanctioning of Law 13.089 Metropolis Statute, we conclude that it was, in a certain way, a great advance to guide public policies in the RMSP. For, through its guidelines, it will guide the joint actions of the federative entities, especially the municipal ones.

CHAPTER 5

BIBLIOGRAPHIC REFERENCES

BRAGA, R., CARVALHO, P. F. **Estatuto da cidade** - política urbana e cidadania. Rio Claro: Laboratório de Planejamento Municipal; DEPLAN-ICGE, UNESP, 2000.

BRAZIL, Ministry of Environment. **CONAMA RESOLUTION No. 001, of 23 January 1986.** Available at :< http://www.mma.gov.br/port/conama/res/res86/res0186.html>. Accessed May 2017.

BRAZIL, Presidency of the Republic. **Constituição da República Federal de 1988**. Disponívelem :< https://www2.senado.leg.br/bdsf/bitstream/handle/id/518231/CF88_Livro_EC9 1_2016.pdf?sequence=1>. Accessed May 2017.

BRAZIL, Presidency of the Republic. Law 13.089; **Metropolis Statute**. Brasília, 2015. Available at: <http://www.planalto.gov.br/ccivil_03/_Ato2015- 2018/2015/Lei/L13089.htm>. Accessed May 2017.

BRAZIL, Presidency of the Republic. Law 10.257/2001; **Statute of the Cities**. Brasília, 2001. Disponívelem : <http://www.planalto.gov.br/ccivil_03/leis/LEIS_2001/L10257.htm> Accessed May 2017.

BRAZIL, Federal Supreme Court. **Direct Action of Unconstitutionality**: ADI1842RJ . Disponívelem : <https://stf.jusbrasil.com.br/jurisprudencia/24807539/acao-direta-de- unconstitutionality-adi-1842-rj-stf>. Accessed: jan. 2017.

CARVALHO, Pompeu Figueiredo. Expert reports and technical opinions on land parcelling and housing construction. In: MAURO, C. A. de. Laudos **periciais em depredações ambientais**. Rio Claro: LPM/DPR/IGCE/UNESP, 1997.

GROSTEIN, MARTA DORA. Metropolis and urban expansion: the persistence of "unsustainable" processes. **Revista Perspectiva.** São Paulo, v. 15, n. 1, p. 1319, Jan. 2001. Disponívelem : <http://www.scielo.br/scielo.php?script=sci_arttext&pid=S0102- 88392001000100003&lng=en&nrm=iso>. Accessed May 2017.

GUZMÁN, J.M.; RODRÍGUEZ, J.; MARTÍNEZ, J. CONTRERAS, J. M.; GONZÁLEZ, D. La démographie de l'Amérique latine et de la Caraibe depuis 1950, **Population** 5/2006 (Vol. 61) , p. 623-733. Available at: < www.cairn.info/revue-population-2006-5-page623.htm.>. Accessed May 2017.

JABAREEN, Y. R. Sustainable urban forms: their typologies, models and concepts. **Journal of Planning Education and Research**, 26: 38-52, 2006.

JENKS, M., JONES, C. **Dimensions of the sustainable city**. London: Springer, 2010.

LEVY, W. Estatuto da Metrópole can correct historical planning omissions. **Revista Uol.** Available at:

<https://noticias.uol.com.br/opiniao/coluna/2015/03/01/estatuto-da-metropole- can-correct-historical-planning-commitments.htm?cmpid>. Accessed: jan. 2017.

MARICATO, Ermínia. As idéias fora do lugar e o lugar fora das idéias: planejamento urbano no Brasil. In: ARANTES, Otília; VAINER. Carlos;
MARICATO, Ermínia. [2000]. **A cidade do pensamento único**: desmanchando consensos. 3ª .
Ed. Petrópolis: Vozes, 2002.

MATOS, R. Aglomerações Urbanas, Redes de Cidades e Desconcentração Demográfica no Brasil. **NEPO,** Unicamp, Campinas, 01 - 21, 2000. Available:
<http://www.abep.nepo.unicamp.br/docs/anais/pdf/2000/Todos/migt4_3.pdf>. Accessed May 2017.

OLIVEIRA, R. F. Regiões metropolitanas e a problemática legal na governança dos serviços públicos de saneamento ambiental no Brasil. **I Congress of Geography and Current Issues**. UNESP - Rio Claro, 2015.

OLIVEIRA, Rafael Fabricio. **De aldeamento jesuítico a periferia metropolitana**: Carapicuíba/SP como rugosidade patrimonial. Thesis (Doctorate in Geography). Brasília: GEA/UNB, 2016.

SABBAGH, R. B. Gestão Ambiental. **Caderno de Gestão Ambiental 16**. State of São Paulo, SMA. São Paulo. 2013. P.101-108.

SILVA, C. M. Brazilian metropolization: political action or socio-spatial dynamics. **Annals of the X ENANPEGE**. p. 872-884. Presidente Prudente: Anpege, 2015.

Printed by Books on Demand GmbH, Norderstedt / Germany